Animal Homes

KINGFISHER

Kingfisher Publications Plc
New Penderel House,
283–288 High Holborn,
London WC1V 7HZ
www.kingfisherpub.com

First published by Kingfisher Publications Plc 2003
This paperback edition printed in 2006
2 4 6 8 10 9 7 5 3
1TR/0706/PROSP/RNB(RNB)/140MA/F
Copyright © Kingfisher Publications Plc 2003

A CIP catalogue record for this book is available from the British Library.

ISBN-13: 978 07534 0857 5
ISBN-10: 0 7534 0857 0

Senior editor: Belinda Weber
Designers: Sam Combes, Joanne Brown
Picture manager: Cee Weston-Baker
Illustrator: Steve Weston
DTP coordinator: Sarah Pfitzner
Artwork archivists: Wendy Allison, Jenny Lord
Senior production controller: Nancy Roberts
Indexer: Chris Bernstein

Printed in Singapore

Acknowledgements
The Publisher would like to thank the following for permission to reproduce their material. Every care has been taken to trace copyright holders. However, if there have been unintentional omissions or failure to trace copyright holders, we apologise and will, if informed, endeavour to make corrections in any future edition.
b = bottom, *c* = centre, *l* = left, *t* = top, *r* = right

Photographs: *cover*: Getty Images; 4–5 National Geographic Image Collection; 8 Oxford Scientific Films (OSF); 9*t* OSF; 9*b* OSF; 10 OSF; 11*t* Getty Images; 11*b* OSF; 12–13 Getty Images; 13*c* Corbis; 14 Corbis; 15*t* Corbis; 15*b* Corbis; 16 Nature Picture Library; 18*tl* Nature Picture Library; 18–19 National Geographic Image Collection; 19*t* Getty Images; 21 Ardea; 22*tr* Nature Picture Library; 22*b* Natural History Picture Agency (NHPA); 23*t* Corbis; 23*b* Nature Picture Library; 26 Corbis; 27 Ardea; 28–29 National Geographic Image Collection; 29*t* Corbis; 30*bl* OSF; 31*t* Corbis; 31*b* Ardea; 32*t* NHPA; 32–33 NHPA; 33*b* NHPA; 35*t* NHPA; 36 Corbis; 36–37 Still Pictures; 37 OSF; 38 Corbis; 38–39 Getty Images; 39 Nature Picture Library

Commissioned photography on pages 42–47 by Andy Crawford.
Thank you to models Eleanor Davis, Lewis Manu, Daniel Newton, Lucy Newton, Nikolas Omilana and Olivia Omilana.

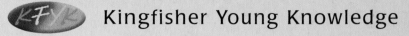
Kingfisher Young Knowledge

Animal Homes

Angela Wilkes

Contents

What is a home?

Animals need homes for all the same reasons as people. Homes provide shelter and keep animals warm in winter. They are a safe place to rest and to bring up babies.

Hard to find

Animals build their homes out of materials that match their surroundings. This makes it hard for predators to spot them.

predators – animals that hunt and eat other animals

Safe place for babies

Homes such as this bird's nest are only built for bringing up babies. A nest is warm, snug and out of the reach of danger. This is where a mother bird lays her eggs and brings up her young.

Different homes

Animals make many kinds of homes. Some build nests, while others make dens, or dig burrows.

Living in a pond

Many different animals live in the still, fresh water of a pond. Here they can find good hiding places, and lots of things to eat.

Finding a mate

Newts live in ponds in the spring. They look for a mate, then lay their eggs in the water.

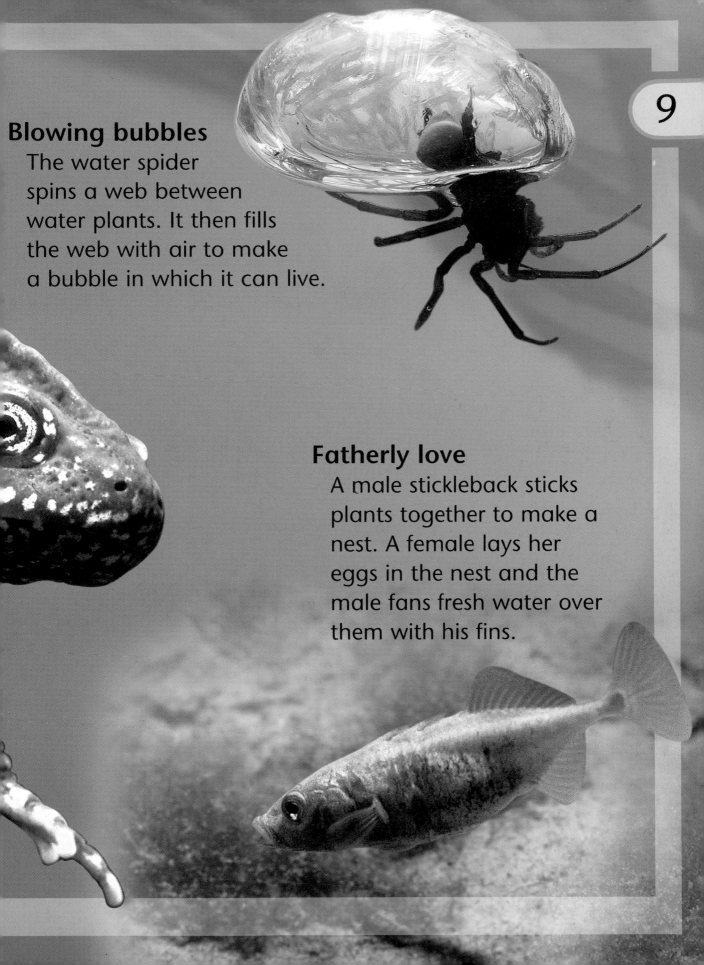

Blowing bubbles

The water spider spins a web between water plants. It then fills the web with air to make a bubble in which it can live.

Fatherly love

A male stickleback sticks plants together to make a nest. A female lays her eggs in the nest and the male fans fresh water over them with his fins.

10 Keeping damp

Amphibians, such as frogs and toads, live in damp, shady places. They need to keep their skin moist and slimy. Some make their homes in unusual spots.

water-holding frog

Living in a hole

When it rains, the water-holding frog's skin soaks up water. The slime on its skin sets to make a cocoon that keeps the water in. The frog then burrows into the sand to escape the desert heat.

amphibian – *animal that is born in water but lives on land later*

Tree-top homes

Strawberry poison-dart frogs live in steamy rainforests. They hide from the hot sun in pools of rainwater that collect in the middle of huge plants.

Burrowing toads

Spadefoot toads dig burrows and spend most of their time there. But when it rains, they come above ground to find a mate.

cocoon – wrapping that protects an animal

Mobile homes

Some animals live in a shell that they carry on their back. The hard shell shields the animal's soft body from knocks and bumps, and shelters it from the wind and rain. It also protects it from hungry predators looking for food.

Body armour

A tortoise has a tough shell shaped like a dome. If the tortoise is in danger, it pulls its head and legs back into its shell.

shields – *protects and looks after*

A home that grows

As a snail grows, its shell grows too, so the shell is always just the right size. Snails slink back inside their shells to hide from danger.

Borrowed home

A hermit crab has no hard shell of its own. So it finds an empty mollusc shell and moves in. When it grows, the hermit crab moves to a bigger shell.

mollusc – an animal with a soft body and a hard shell

Spinning webs

Most spiders spin webs to catch insects. They build them out of silk threads from their own bodies. But some spiders live in other types of home, such as holes or burrows.

Cobweb trap

The orb spider spins a beautiful sticky web between the stems of plants. The spider lies in wait for insects in the centre of the web, or hides under a nearby leaf.

prey – an animal that is hunted and killed by another animal

Under a rock

Some spiders make nests in hollows under rocks. They line the nest with thick silk and lay their eggs. Then they wait to pounce on passing insects.

Ambush!

The trapdoor spider digs a tunnel and lines it with silk. Then it makes a lid on top, like a trapdoor. The spider hides in the tunnel and darts out to catch prey.

Where birds live

Birds build nests so they have a warm, safe place to lay their eggs and look after their chicks. Most birds' nests are in trees, but some are on steep cliffs or even on the ground.

Hanging nest

The penduline tit hangs its purse-shaped nest from a twig. It is lined with soft wool and fluff from reeds or catkins to make a cosy home for the female and babies.

catkin – a spike of small, soft flowers on a tree

Fancy work

A male cape weaver bird weaves a complicated oval nest out of grass and reeds. The only way in is through a short tunnel at the bottom. This helps protect it from predators.

Hungry chicks

Chicks hatch in their nest. They are helpless and cannot leave. They open their beaks wide to beg for food.

*wasp larva
in a cell*

Cell homes

Wasps and bees build fantastic nests, made up of lots of tiny cells. Young wasps and bees can grow up safely in these little compartments.

Laying eggs

A queen wasp lays one egg in each cell. Each egg will become a wasp larva. Older wasps look after the eggs and larvae.

Paper home

A wasp's nest is made of layers of paper around the larvae cells. The wasps make paper by chewing up wood and mixing it with their spit.

larva – young insect that has just come out of its egg

Moving house

Bees live and work together. When bees need a new home, they fly away in a huge group called a swarm.

Honeycomb homes

Bees' nests are called hives. Inside a hive are wax honeycombs made of lots of cells. The cells hold honey or a baby bee.

larvae – more than one larva

leafcutter ants

Living in a colony

Most ants and termites live in huge groups called colonies. They work together to build enormous nests in which to raise their young.

Food for the colony

Leafcutter ants live in rainforests. They bite off pieces of leaf and carry them back to their nest. They store the leaves in special gardens and a fungus grows on them, making a tasty food for the ants.

queen termite

fungus – *a plant, such as a mushroom, that grows on other plants*

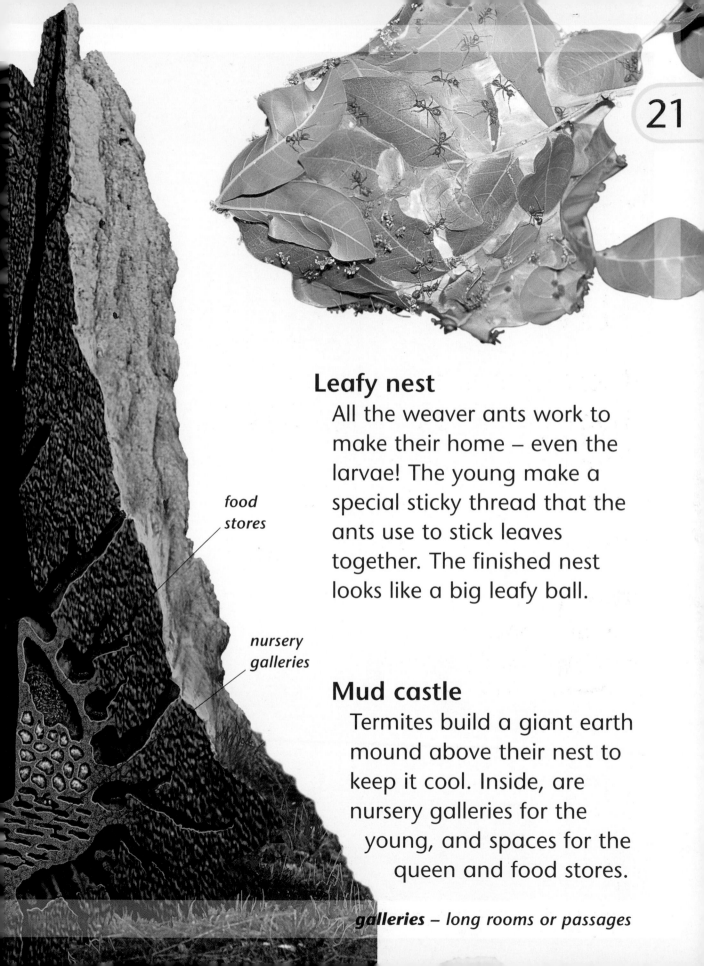

food
stores

nursery
galleries

Leafy nest

All the weaver ants work to make their home – even the larvae! The young make a special sticky thread that the ants use to stick leaves together. The finished nest looks like a big leafy ball.

Mud castle

Termites build a giant earth mound above their nest to keep it cool. Inside, are nursery galleries for the young, and spaces for the queen and food stores.

galleries – long rooms or passages

Mini homes

Mice live in lots of places. Some live in fields and some live in woods. Others even live in people's houses. But all mice build nests to rest in and bring up their babies.

Close to people

House mice make their nests from shredded paper, old rags or grass. They always build them in a small hiding place well out of sight.

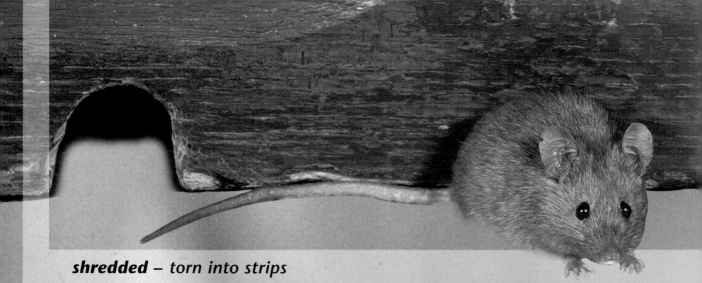

shredded – torn into strips

Grassy homes

The tiny harvest mouse lives in tall grass. It weaves strips of grass around plant stems to make a snug, round nest.

Sleepy mouse

In the autumn, the dormouse makes a cosy nest out of shredded bark. Then it curls up into a ball and goes to sleep for the long, cold winter.

bark – *the outer layer of a tree's trunk or branches*

Under the ground

Marmots live high in the mountains. When the first winter snows fall, the whole family moves into a big burrow lined with grass, and goes to sleep.

Fast asleep

The marmots block the entrance to the burrow with rocks and soil to stop predators getting in. Then they snuggle together for warmth, and hibernate until spring comes again.

hibernate – spend the whole winter in a deep sleep

Where dogs live

Foxes and dingoes are wild dogs. They make their homes by digging dens in soft earth, or by taking over and enlarging the homes of other animals. Dens provide shelter from the hot sun or cold weather, and are a safe place to bring up pups.

Desert homes
Kit foxes live in stony deserts in North America. They sleep in their dens during the day, when the sun is at its hottest. At night, when it is cooler, they go out hunting.

den – *a sheltered place where an animal lives*

Ready-made den

Dingoes live in Australia. When a mother dingo is about to have pups, she moves into a safe den. This is often a big hole beneath tree roots or some rocks.

enlarging – *making bigger*

28 Living in the snow

Polar bears live in the snowy Arctic. When they are tired, they dig a shallow pit in the snow, and sleep in it. In the autumn, a pregnant polar bear digs a den in a snowdrift. This is where she will spend the long, dark winter.

Snow babies

The mother bear stays in the den until the spring. In the early winter, she gives birth to one or two cubs. She feeds them her milk, and they all sleep for most of the winter.

pregnant – *going to have a baby*

Leaving the den

In the spring, the polar bear and young come out of the den. The mother is very hungry as she has not eaten all winter. She takes her cubs on to the sea ice, where she can hunt for food.

sea ice – *ice that forms on the surface of the sea when it freezes*

Going batty

Bats go out hunting for food at night and rest during the day. They do not make special homes, but roost in trees, caves, barns or even attics.

Leafy shelter

Fruit bats live in huge groups called colonies and roost in tall trees by day. They hang upside down from branches, clinging on tightly with the claws on their feet. Then they wrap their skinny wings around themselves for protection.

roost – *settle down to sleep*

Dark caves

Many bats sleep in large caves.
Thousands of them roost upside down,
packed closely together. When
evening comes, the bats set out
to feed. Some bats feed on
insects. Others, like these
flying foxes, eat fruit and
the nectar from flowers.

dragonfly

Riverside
homes

Many animals live on the banks of rivers and streams. Here, they are close to fresh water and there are plenty of plants, small creatures and fish to eat. They are also safely out of the reach of most predators.

Water babies

Adult dragonflies live beside rivers. The larvae live in the water. When the larvae are ready to become adults, they climb up a plant's stem. Their skin splits open along their backs, and the adult dragonfly climbs out.

Nest tunnel

Kingfishers dig a tunnel in a soft river bank. At the end of the tunnel the female kingfisher makes a small chamber and lays her eggs. When the chicks hatch, she brings them fish to eat.

Rest burrows

Platypuses live near lakes and rivers. A mother platypus digs a long, nesting burrow in the soft earth of the bank. Here, she lays her eggs and looks after her babies.

chamber – a room

Living in a lodge

Beavers are clever builders. They construct dams across streams to make ponds. Then they build homes, called lodges, in the middle of these ponds.

Safe from enemies

The beavers line the lodge with dry grasses to keep it snug and warm. All the entrances are under water, safe from predators.

Timber!

Using their sharp front teeth, beavers can cut down trees. They gnaw around the bottom of a tree until it falls down. Then they chew pieces off to make small logs.

Saving food for later

Beavers only eat plants. They store some food at the bottom of the pond, so they can eat all winter.

Going camping

Big apes, such as chimpanzees, orang-utans and gorillas, do not have one home. They move from place to place. At night, they make leafy nests and camp out.

Climbing trees

Chimpanzees make tree nests at night-time. They bend leafy branches over to make comfortable beds on which they can sleep.

ape – an animal that looks like a monkey without a tail

Leafy nests

Orang-utans make two tree nests a day. They make a small nest for a nap. At night, they build platforms in the forks of trees.

Heavy sleepers

Female gorillas nest in trees. Male gorillas make nests on the ground as they are too heavy to sleep in trees!

Living with people

Towns and cities offer shelter, food and warmth. As they have grown in size, more animals have moved into them. Animals often settle into new homes in the most surprising places.

Cardboard bed

In North America, raccoons have moved to town gardens and even into city centres. They live in attics and sheds. Raccoons eat almost anything, and even help themselves to food from dustbins.

platform – flat surface

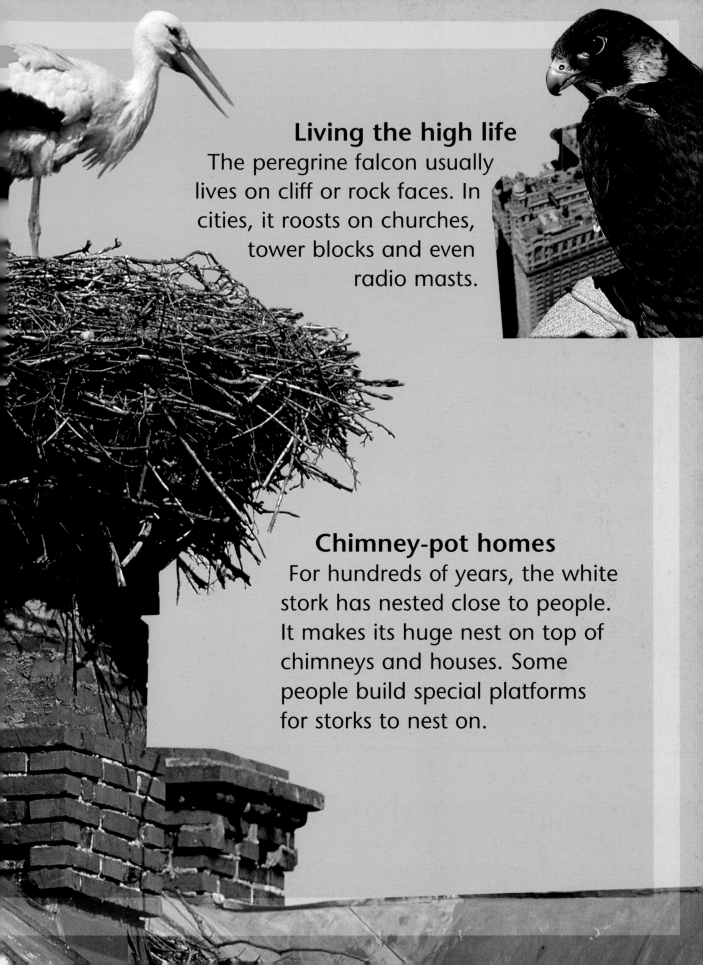

Living the high life

The peregrine falcon usually lives on cliff or rock faces. In cities, it roosts on churches, tower blocks and even radio masts.

Chimney-pot homes

For hundreds of years, the white stork has nested close to people. It makes its huge nest on top of chimneys and houses. Some people build special platforms for storks to nest on.

40 Looking at homes

You will need
- Plastic cup
- Scissors
- Plastic food wrap
- Elastic bands

Make a pond viewer
With this simple underwater viewer you can take a closer look at the small creatures that live in ponds and streams.

Hold the plastic cup firmly in one hand. Then hold the scissors in your other hand and carefully cut out the bottom of the cup.

Cut out a large circle of plastic food wrap. Stretch it tightly over the cut end of the cup, until the food wrap is smooth.

Stretch a couple of elastic bands over the food wrap to hold it in place. Pull the edges of the plastic food wrap tight again.

To use the pond viewer, dip the end covered with food wrap into the water. Then look through the open end at the top of the cup.

Make a nest

Watch different birds making their nests in spring. See if you can copy them by making a bird's nest of your own.

You will need

- Paintbrush and glue
- Plastic bowl
- Dried grass
- Moss
- Feathers and leaves
- Sweet wrappers

1

Using a paintbrush, spread glue all around the outside of the bowl. Pick up small handfuls of dried grass and stick them to the bowl.

2

Spread glue around the inside of the bowl. Then stick on a layer of more dried grass and moss, to make a soft, cosy middle.

3

Scatter a few small feathers and leaves inside the nest to make it look realistic. Decorate with a few sweet wrappers to add colour.

42 Watching animals

Make a teepee hide

If you make a simple teepee in your garden or local park, you can hide inside it and watch animals.

You will need
- Large sheet
- 4 bamboo canes
- Garden twine
- Paintbrush
- Scissors
- Poster paints

Mix the paints with a little water, so they are easy to use. Paint circles and other bold shapes on the sheet. Leave the sheet to dry.

Stand the four bamboo canes together and make them into a teepee shape. Cut a long piece of twine and tie the canes together firmly at the top.

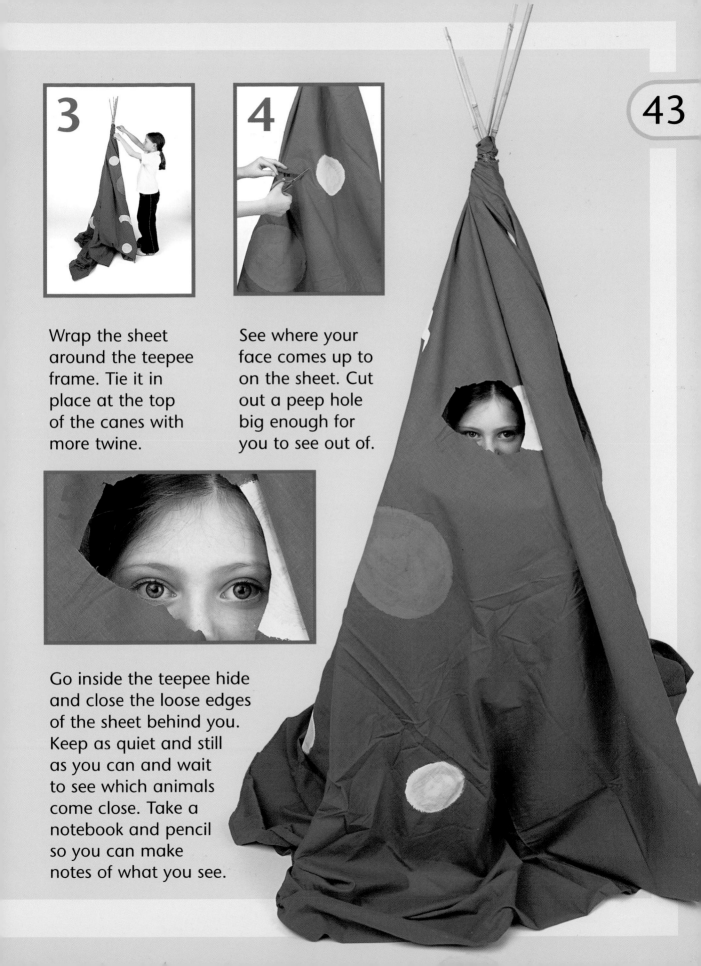

3

Wrap the sheet around the teepee frame. Tie it in place at the top of the canes with more twine.

4

See where your face comes up to on the sheet. Cut out a peep hole big enough for you to see out of.

Go inside the teepee hide and close the loose edges of the sheet behind you. Keep as quiet and still as you can and wait to see which animals come close. Take a notebook and pencil so you can make notes of what you see.

44 Making homes

Make a hermit crab

Make a crab out of modelling dough and put it in an empty shell. Then it will be just like a real hermit crab.

You will need
- Shell
- Modelling dough
- Pipe-cleaner

Roll two pieces of modelling clay into balls for the crab's head and body. Make four small sausages for legs and two claw shapes.

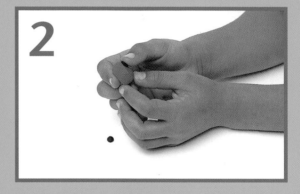

Break off two tiny pieces of a different colour modelling clay. Roll them into little balls for the eyes, and stick them on to the head.

Stick the head, claws and legs to the crab's body. Make feelers from two bits of pipe-cleaner, then put the crab in the shell.

Flowerpot home

Make a flowerpot home for minibeasts. Check to see what is inside it every day, and draw the creatures you find there.

You will need
- Flowerpot
- Small rock
- Notebook and pen

Ask an adult to help you find a shady spot somewhere near your home. Turn the flowerpot upside down and prop it up on the rock.

After a few days, look inside the flowerpot. Draw pictures in your notebook of any creatures you find. Can you name them?

Bee home

Make this simple bee box and hang it in a sunny place outside. The straws should slope down into the bottle.

You will need
- Large plastic bottle
- Drinking straws
- Scissors
- String

Cut the top end off the bottle, and fill it with straws. Then tie string around the bottle and hang it outside.

46 Hamster funbox

Build a funbox

Your pet hamster or mouse will have lots of fun with this play box. It can climb in and out of it, as if on a climbing frame.

You will need
- Shoebox
- 4 cardboard tubes
- Poster paints
- Scissors
- Paintbrush
- Pencil
- Cup

Position a cardboard tube in the middle of one end of the shoebox and draw round it. Repeat on the other sides of the box.

Make a hole in the centre of a circle and cut lines out to the edge of it. Then cut out the circle. Do this with all the circles on the box.

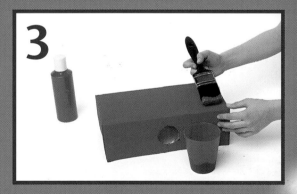

Mix some poster paint with a little water and carefully paint the cardboard box all over. Then leave the paint to dry.

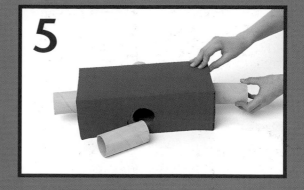

Paint the cardboard tubes a different colour. Paint one end of each tube and let it dry. Then paint the second half of the tube and leave that to dry too.

Push the cardboard tubes into the holes around the side of the box. They should fit firmly and stick out a bit. Now see if your hamster wants to play.

Index